Curaçao, 2023

This book is written / compiled by
John H Baselmans-Oracle

Layout and front by
John H Baselmans-Oracle

With special thanks to those people who have helped me.

Copyrights

Included all writers who have contributed to
this anthology.
This reference is to show from the widest possible spectrum,
how the matrix of this system works.
Personally I do not believe in these rights because
as a free man of flesh and blood
you own nothing.

Note:
The photos you see in chapter 1 were found via the internet.
Unfortunately, no references or texts have been found let alone the creators of these photos.
But if there is any mention of them somewhere,
I would certainly be very happy to include them with the photos.
Thank you for your cooperation.
John

ISBN 978-1-312-54931-9

Quantiversum Portal
Past and Now

INDEX

Quantiversum 5

Part 1
The old age.
THE GATES AND PORTALS DID EXIST 6
Present day 41

Part 2
Quantiverse Portal Curacao 49
DESCRIPTION FRONT 58
DESCRIPTION BACK 70
Frequency Quativersum Portal 75

According to ancient writings, man's life is said to be
determined by one act of a man and a woman.
The "time" factor is also what people subsequently live by.

*

Unfortunately, this kind of errors
have been going on for as long as humanity has existed.

*

Despite the true path of life being visible,
people do not want to acknowledge it let alone follow it.

*

The Quantiversum portal is nothing new.
It is a journey to the core of life.
A technique that has existed from day one in humanity.

*

The Quantiversum portal in Curaçao is unique!
However the question remains; can people handle this journey?

*

The journey in a true life, in other words, the journey to the core of energy!

Curaçao, May 2022 - January 2023

Part 1
The old age

THE GATES AND PORTALS DID EXIST

People from the continents of the Earth were able to relocate to other worlds/earths by ships. It's not for nothing that they say space elf, because they were real ships sailing through the gates in ice to other worlds and not to the empty space above Earth.

Universe means "Whole World" in Czech.

People sailed through the gates or to travel to the Whole World used Portals, based on a similar system used in the Philadelphia Experiment. Some places in the world like the island of Rapa-Iti, still have the Portal.

Portals were either rebuilt or demolished.

Where all the old portals/gates have gone nobody knows and apart from several having been destroyed, there must still be these portals in various places. Apparently, people do not yet understand how it works. It all seems that we are not allowed to know anything about these working portals.

From many of these "disappeared" portals, I have placed photos in this book to show that there is more out there than we think we can see. In the second part of the book, I show the construction of the world's first Quantiversum Portal and built on the island of Curaçao.

While locally, people do not yet understand much of this sculpture and most humankind are timid and afraid. However, there is already international attention and questions being asked that go deep into life.

Please welcome to the real world that is revealing the meaning of life.

Portal

Portal

Portal

Portal

Portal

Portal

Portal

Portal

Portal

Portal

Portal

Portal

Portal

Portal

Portal

Portal

Portal

Present day

Portal

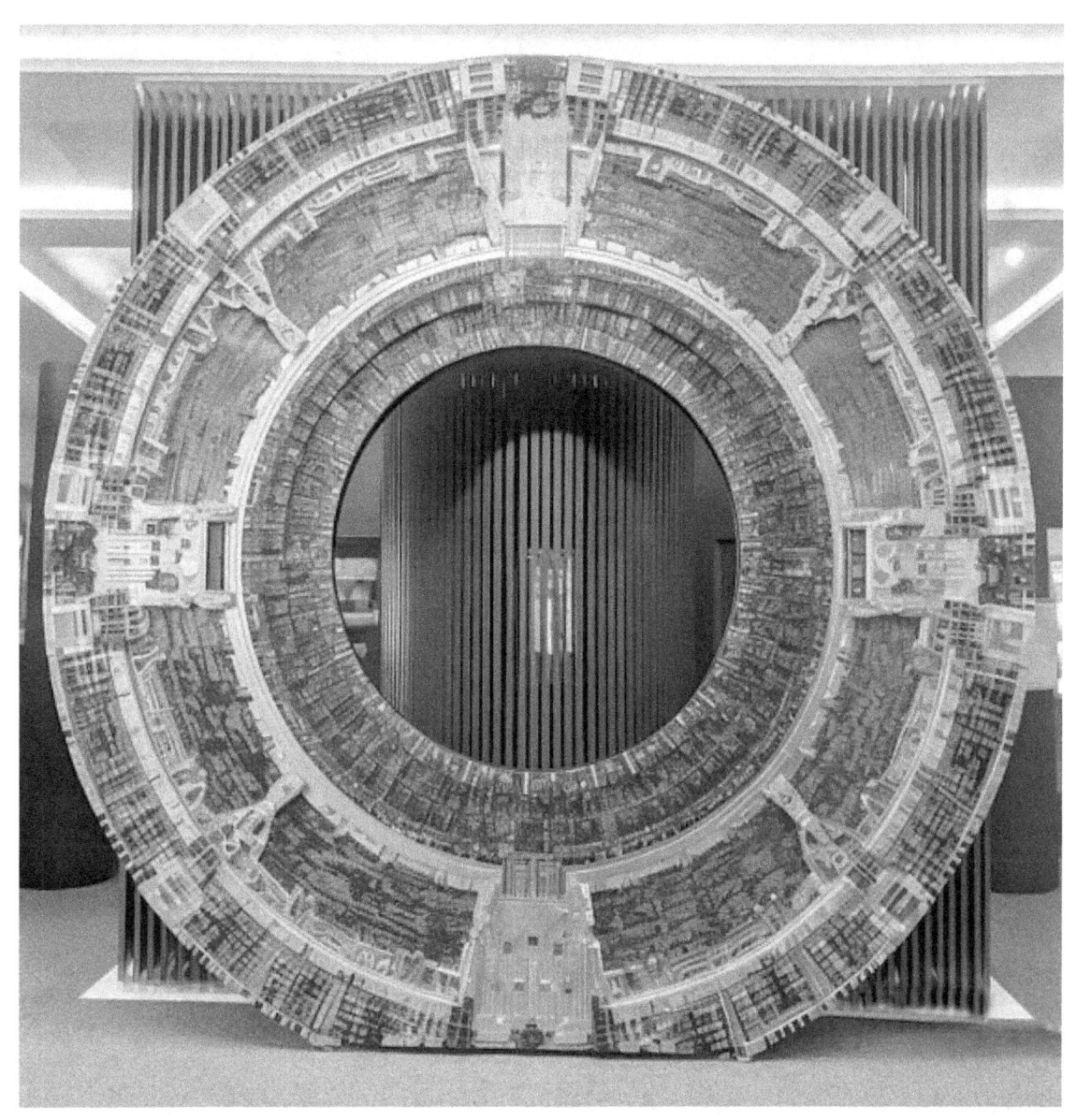

Portal

Portal

Part 2
Quantiverse Portal Curacao

In 2019, the idea of a "Quantiversum Portal", which is the connection to all the dimensions out there, was born. It had started with a concept of making a big round object which was to be some kind of gate/portal. Years before, I had started with a "formwork" to make the elements in concrete. Right at the beginning, it turned out that the idea was failing as I just couldn't work with those elements due to the immense weight.

The first setup had not got very far as making it out of concrete with a steel structure for strength was a hopeless task. It would become difficult and too heavy to accomplish this alone. Consequently, this stopped the idea and what also turned out later was that the first plan had many gaps.

It wasn't until 2022 that the plan revived and I got the idea to make the whole thing out of fibre. I started making drawings of the structure that had to hold it all. It had to be a sturdy inner structure, not only for the 400-centimetre-diameter, 50-centimetre-wide colossus but it also had to be able to withstand the local gust. The carcass was made of 10- and 12-millimetre-thick rebar.

With the frame inside made of 20 by 50 millimetre box section and the base made of 50 m/m angle iron and box section. A lightning rod was incorporated into the whole as the 4-metre-high structure is going to be mostly steel and also is located at the highest point of our garden.

To fibre the whole thing reasonably evenly, I wanted to have coarse concrete effect, I used cardboard sheets covered with at least 4 layers of fibre as a substrate. Which were all installed as a circle and on which all the symbols were then applied afterwards.

One of the biggest obstacles is my fear of heights! 2 steps on a staircase already makes me insecure. But I had found a way to assemble the many elements. Because of the fear of heights, I used the set-up of making the highest quarter section completely ready and then assembling on the highest point of the whole only. This is how I worked until I got to the bottom part and could finish everything.

Furthermore, making 80 symbols has been a long road with many setbacks where the plaster kept resisting, that along with frequent rainfall and sometimes severe weather.

A very remarkable thing is that as I started working with the symbols, I started feeling sicker and sicker. It turned out that somewhere the symbols had a tremendously heavy energy around them that made working with them increasingly difficult for me. Throughout the time I was making the symbols, there had been severe setbacks. There were symbols among them that I had to remake up to five times and even broke down during the drying period. The one of the big things was some mega lightning strikes that destroyed many electrical and electronic equipment. Even my wife noticed that since starting symbol making, many things did go wrong and both of us felt sick.

It was clear that I had to deal with strange forces even before the project started. Something I did take into account and that started to make itself known while building this portal. Eventually, by hanging 4 iron balls inside, I got a better grip on this immeasurable energy. Later, other components were built in to regulate the energy.

In the final stages of building this sculpture, it was fighting complete exhaustion. The energy and forces that were all going their own way may have been bundled but I was working in them daily. It was only when the symbols were put in place that one whole was created and everything came into balance. The stress of whether it will all work out remained a question until the last day. It is playing with so many energies that the architect of life works with and is unknown territory for humankind. It is and always will be a mystery to us humankind. Many humankind are afraid and don't know how to deal with all this power.

The colossus has been placed in a column of concrete in the ground and it is now working and giving us an insight into the world as it really is.

Should you ever visit the area, come and see and feel for yourself the true forces of life.

Welcome John

Portal

Portal

Portal

"Quantiversum Portal" is working

It was in October 2022 that I announced the
world premiere of the first "Quantiversum Portal".

And as expected; the art world, press and politics were the big ones absent.
This is quite understandable because the eight-month
project with over 1,500 hours of work
and many calculations was done without grant money, nor donations
let alone any support from outside. Everything (many thousands of guilders)
was paid for out of our own pockets!
In fact, I had to devise, calculate and execute the entire project myself.
Working 4 meters high and that with vertigo and the necessary complications.

The sculpture reported from the very beginning that
the present time was not to be created with the future and the past.
Despite all the obstacles, the project was completed and, as all locations
were rejected for years now, it now stands in its full glory in our own garden!

The "Quantiversum Portal" is up and running and strange things
are happening of which I have reported several times.

The entire construction and work on the sculpture will soon be shown and
read in detail on my website. Because of the disinterest locally, it will not be
accessible to those not really interested as Curaçao has made it clear that
a world first is NOT of any interest if no money is shoved under the table.

"Quantiversum Portal" is a reality and I have proven that a mega-project
with heavy and large elements can be achieved by a single person
with the support of his wife. "Quantiversum Portal" is now visible and working
for humankind, and that in a small island to be called Curaçao.
Curaçao, May 2022 - January 2023

DESCRIPTION FRONT

Glyphs are symbols on Stargates which chevrons lock onto when a Stargate is being dialed. The basis for glyphs are star constellations. There are several differences between Milky Way and Pegasus galaxy glyphs.

At the beginning, that the Symbol on the Stargate, which represent the Earth is combine of a triangle and a circle because of the Stargate was found at the pyramids. So the triangle represent the pyramids und the circle the stargate. But after all we know, the Goa'uld stole this technology from the ancients. So my point is, when the pyramids where build, the Stargates already existed. So there would be no point in having a triangle representing a pyramid.

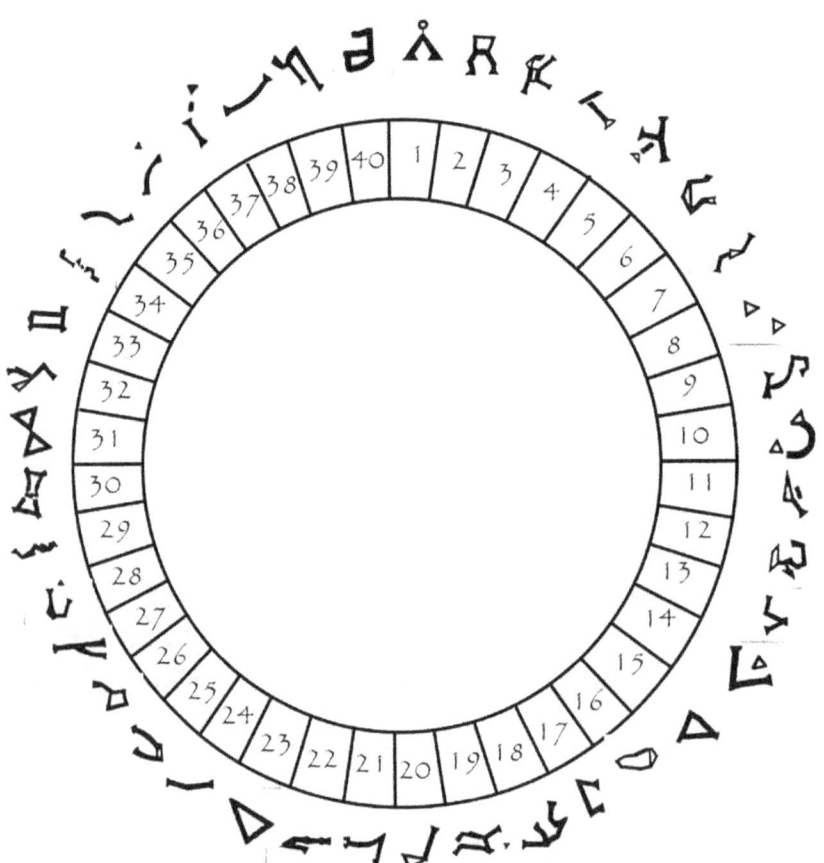

1
Point of origin

The "point of origin" symbol for Earths Alpha Gate, called "At" in Ancient.

The point of origin is a symbol necessary for the dialing of Stargate addresses. While the first six symbols represent constellations indicating a point in space, this seventh point indicates where the wormhole is coming from. Milky Way and Pegasus Stargates each have a unique point of origin symbol, easily distinguished from the rest by those familiar with the technology. Destiny-style Stargates have a uniform design, and thus the point of origin can be any of the 36 symbols on the gate, depending on where it is dialed from.

2

Crater is a small constellation in the southern celestial hemisphere. Its name is the latinization of the Greek krater, a type of cup used to water down wine. One of the 48 constellations listed by the second-century astronomer Ptolemy, it depicts a cup that has been associated with the god Apollo and is perched on the back of Hydra the water snake.

There is no star brighter than third magnitude in the constellation. Its two brightest stars, Delta Crateris of magnitude 3.56 and Alpha Crateris of magnitude 4.07, are ageing orange giant stars that are cooler and larger than the Sun. Beta Crateris is a binary star system composed of a white giant star and a white dwarf. Seven star systems have been found to host planets. A few notable galaxies, including Crater 2 and NGC 3981, and a famous quasar lie within the borders of the constellation

3

Virgo is one of the constellations of the zodiac. Its name is Latin for maiden, and its old astronomical symbol is Virgo symbol. Lying between Leo to the west and Libra to the east, it is the second-largest constellation in the sky (after Hydra) and the largest constellation in the zodiac. The ecliptic intersects the celestial equator within this constellation and Pisces. Underlying these technical two definitions, the sun passes directly overhead of the equator, within this constellation, at the September equinox. Virgo can be easily found through its brightest star, Spica.4 Boötes (/bo?'o?ti?z/ boh-OH-teez) is a constellation in the northern sky, located between 0° and +60° declination, and 13 and 16 hours of right ascension on the celestial sphere. The name comes from Latin: Bootes, which comes from Greek: ???t??, translit. Bo?tes 'herdsman' or 'plowman' (literally, 'ox-driver'; from ß??? boûs 'cow').

One of the 48 constellations described by the 2nd-century astronomer Ptolemy, Boötes is now one of the 88 modern constellations. It contains the fourth-brightest star in the night sky, the orange giant Arcturus. Epsilon Boötis, or Izar, is a colourful multiple star popular with amateur astronomers. Boötes is home to many other bright stars, including eight above the fourth magnitude and an additional 21 above the fifth magnitude, making a total of 29 stars easily visible to the naked eye.

5

Centaurus /s?n't??r?s, -'t??r-/ is a bright constellation in the southern sky. One of the largest constellations, Centaurus was included among the 48 constellations listed by the 2nd-century astronomer Ptolemy, and it remains one of the 88 modern constellations. In Greek mythology, Centaurus represents a centaur; a creature that is half human, half horse (another constellation named after a centaur is one from the zodiac: Sagittarius). Notable stars include Alpha Centauri, the nearest star system to the Solar System, its neighbour in the sky Beta Centauri, and V766 Centauri, one of the largest stars yet discovered. The constellation also contains Omega Centauri, the brightest globular cluster as visible from Earth and the largest identified in the Milky Way, possibly a remnant of a dwarf galaxy.

6

Libra /'li?br?/ is a constellation of the zodiac and is located in the Southern celestial hemisphere. Its name is Latin for weighing scales. Its old astronomical symbol is Libra symbol. It is fairly faint, with no first magnitude stars, and lies between Virgo to the west and Scorpius to the east. Beta Librae, also known as Zubeneschamali, is the brightest star in the constellation. Three star systems are known to have planets.

7

Serpens (Ancient Greek: ?f??, romanized: Óphis, lit. 'the Serpent') is a constellation in the northern celestial hemisphere. One of the 48 constellations listed by the 2nd-century astronomer Ptolemy, it remains one of the 88 modern constellations designated by the International Astronomical Union. It is unique among the modern constellations in being split into two non-contiguous parts, Serpens Caput (Serpent Head) to the west and Serpens Cauda (Serpent Tail) to the east. Between these two halves lies the constellation of Ophiuchus, the "Serpent-Bearer". In figurative representations, the body of the serpent is represented as passing behind Ophiuchus between Mu Serpentis in Serpens Caput and Nu Serpentis in Serpens Cauda.

The brightest star in Serpens is the red giant star Alpha Serpentis, or Unukalhai, in Serpens Caput, with an apparent magnitude of 2.63. Also located in Serpens Caput are the naked-eye globular cluster Messier 5 and the naked-eye variables R Serpentis and Tau4 Serpentis. Notable extragalactic objects include Seyfert's Sextet, one of the densest galaxy clusters known; Arp 220, the prototypical ultraluminous infrared galaxy; and Hoag's Object, the most famous of the very rare class of galaxies known as ring galaxies.

8

Norma is a small constellation in the Southern Celestial Hemisphere between Ara and Lupus, one of twelve drawn up in the 18th century by French astronomer Nicolas-Louis de Lacaille and one of several depicting scientific instruments. Its name is Latin for normal, referring to a right angle, and is variously considered to represent a rule, a carpenter's square, a set square or a level. It remains one of the 88 modern constellations.

Four of Norma's brighter stars—Gamma, Delta, Epsilon and Eta—make up a square in the field of faint stars. Gamma2 Normae is the brightest star with an apparent magnitude of 4.0. Mu Normae is one of the most luminous stars known, with a luminosity between a quarter million and one million times that of the Sun. Four star systems are known to harbour planets. The Milky Way passes through Norma, and the constellation contains eight open clusters visible to observers with binoculars. The constellation also hosts Abell 3627, also called the Norma Cluster, one of the most massive galaxy clusters known.

9

Scorpius is a zodiac constellation located in the Southern celestial hemisphere, where it sits near the center of the Milky Way, between Libra to the west and Sagittarius to the east. Scorpius is an ancient constellation that pre-dates the Greeks; it is one of the 48 constellations identified by the Greek astronomer Ptolemy in the second century. Its old astronomical symbol is Scorpius symbol

10

Corona Australis is a constellation in the Southern Celestial Hemisphere. Its Latin name means "southern crown", and it is the southern counterpart of Corona Borealis, the northern crown. It is one of the 48 constellations listed by the 2nd-century astronomer Ptolemy, and it remains one of the 88 modern constellations. The Ancient Greeks saw Corona Australis as a wreath rather than a crown and associated it with Sagittarius or Centaurus. Other cultures have likened the pattern to a turtle, ostrich nest, a tent, or even a hut belonging to a rock hyrax.

Although fainter than its northern counterpart, the oval- or horseshoe-shaped pattern of its brighter stars renders it distinctive. Alpha and Beta Coronae Australis are the two brightest stars with an apparent magnitude of around 4.1. Epsilon Coronae Australis is the brightest example of a W Ursae Majoris variable in the southern sky. Lying alongside the Milky Way, Corona Australis contains one of the closest star-forming regions to the Solar System—a dusty dark nebula known as the Corona Australis Molecular Cloud, lying about 430 light years away. Within it are stars at the earliest stages of their lifespan. The variable stars R and TY Coronae Australis light up parts of the nebula, which varies in brightness accordingly.

11

The scutum (Classical Latin: ˈskuːtʊ̃ː]; plural scuta) was a type of shield used among Italic peoples in antiquity, most notably by the army of ancient Rome starting about the fourth century BC.

The Romans adopted it when they switched from the military formation of the hoplite phalanx of the Greeks to the formation with maniples (Latin: manipuli). In the former, the soldiers carried a round shield, which the Romans called a clipeus. In the latter, they used the scutum, which was larger. Originally it was oblong and convex, but by the first century BC it had developed into the rectangular, semi-cylindrical shield that is popularly associated with the scutum in modern times. This was not the only kind the Romans used; Roman shields were of varying types depending on the role of the soldier who carried it. Oval, circular and rectangular shapes were used throughout Roman history.

12

Sagittarius is one of the constellations of the zodiac and is located in the Southern celestial hemisphere. It is one of the 48 constellations listed by the 2nd-century astronomer Ptolemy and remains one of the 88 modern constellations. Its old astronomical symbol is Sagittarius symbol . Its name is Latin for "archer". Sagittarius is commonly represented as a centaur pulling back a bow. It lies between Scorpius and Ophiuchus to the west and Capricornus and Microscopium to the east.

The center of the Milky Way lies in the westernmost part of Sagittarius.

13

Aquila is a constellation on the celestial equator. Its name is Latin for 'eagle' and it represents the bird that carried Zeus/Jupiter's thunderbolts in Greek-Roman mythology.

Its brightest star, Altair, is one vertex of the Summer Triangle asterism. The constellation is best seen in the northern summer, as it is located along the Milky Way. Because of this location, many clusters and nebulae are found within its borders, but they are dim and galaxies are few.

14

Microscopium /?ma?kr?'sk?pi?m/ ("the Microscope") is a minor constellation in the southern celestial hemisphere, one of twelve created in the 18th century by French astronomer Nicolas-Louis de Lacaille and one of several depicting scientific instruments. The name is a Latinised form of the Greek word for microscope. Its stars are faint and hardly visible from most of the non-tropical Northern Hemisphere.

The constellation's brightest star is Gamma Microscopii of apparent magnitude 4.68, a yellow giant 2.5 times the Sun's mass located 223 ± 8 light-years distant. It passed within 1.14 and 3.45 light-years of the Sun some 3.9 million years ago, possibly disturbing the outer Solar System. Two star systems—WASP-7 and HD 205739—have been determined to have planets, while two others—the young red dwarf star AU Microscopii and the sunlike HD 202628—have debris disks. AU Microscopii and the binary red dwarf system AT Microscopii are probably a wide triple system and members of the Beta Pictoris moving group. Nicknamed "Speedy Mic", BO Microscopii is a star with an extremely fast rotation period of 9 hours, 7 minutes.

15

Capricornus /?kæpr?'k??rn?s/ is one of the constellations of the zodiac. Its name is Latin for "horned goat" or "goat horn" or "having horns like a goat's", and it is commonly represented in the form of a sea goat: a mythical creature that is half goat, half fish.

Capricornus is one of the 88 modern constellations, and was also one of the 48 constellations listed by the 2nd century astronomer Claudius Ptolemy. Its old astronomical symbol is Capricornus symbol. Under its modern boundaries it is bordered by Aquila, Sagittarius, Microscopium, Piscis Austrinus, and Aquarius. The constellation is located in an area of sky called the Sea or the Water, consisting of many water-related constellations such as Aquarius, Pisces and Eridanus. It is the smallest constellation in the zodiac.

16

Piscis Austrinus is a constellation in the southern celestial hemisphere. The name is Latin for "the southern fish", in contrast with the larger constellation Pisces, which represents a pair of fish. Before the 20th century, it was also known as Piscis Notius. Piscis Austrinus was one of the 48 constellations listed by the 2nd-century astronomer Ptolemy, and it remains one of the 88 modern constellations. The stars of the modern constellation Grus once formed the "tail" of Piscis Austrinus. In 1597 (or 1598), Petrus Plancius carved out a separate constellation and named it after the crane.

It is a faint constellation, containing only one star brighter than 4th magnitude: Fomalhaut, which is 1st magnitude and the 18th-brightest star in the night sky. Fomalhaut is surrounded by a circumstellar disk, and possibly hosts a planet. Other objects contained within the boundaries of the constellation include Lacaille 9352, the brightest red dwarf star in the night sky (though still too faint to see with the naked eye); and PKS 2155-304, a BL Lacertae object that is one of the optically brightest blazars in the sky.

17

Equuleus (/?'kwu?li?s/ ih-KWOO-lee-?s) is a constellation of stars that are visible in the night sky. Its name is Latin for "little horse", a foal. Located just north of the celestial equator, it was one of the 48 constellations listed by the 2nd century astronomer Ptolemy, and remains one of the 88 modern constellations. It is the second smallest of the modern constellations (after Crux), spanning only 72 square degrees. It is also very faint, having no stars brighter than the fourth magnitude.

18

Aquarius is an equatorial constellation of the zodiac, between Capricornus and Pisces. Its name is Latin for "water-carrier" or "cup-carrier", and its old astronomical symbol is Aquarius symbol , a representation of water. Aquarius is one of the oldest of the recognized constellations along the zodiac (the Sun's apparent path).2] It was one of the 48 constellations listed by the 2nd century astronomer Ptolemy, and it remains one of the 88 modern constellations. It is found in a region often called the Sea due to its profusion of constellations with watery associations such as Cetus the whale, Pisces the fish, and Eridanus the river.3]

At apparent magnitude 2.9, Beta Aquarii is the brightest star in the constellation.

19

Pegasus is a constellation in the northern sky, named after the winged horse Pegasus in Greek mythology. It was one of the 48 constellations listed by the 2nd-century astronomer Ptolemy, and is one of the 88 constellations recognised today.

With an apparent magnitude varying between 2.37 and 2.45, the brightest star in Pegasus is the orange supergiant Epsilon Pegasi, also known as Enif, which marks the horse's muzzle. Alpha (Markab), Beta (Scheat), and Gamma (Algenib), together with Alpha Andromedae (Alpheratz) form the large asterism known as the Square of Pegasus. Twelve star systems have been found to have exoplanets. 51 Pegasi was the first Sun-like star discovered to have an exoplanet companion.

20

Sculptor is a small and faint constellation in the southern sky. It represents a sculptor. It was introduced by Nicolas Louis de Lacaille in the 18th century. He originally named it Apparatus Sculptoris (the sculptor's studio), but the name was later shortened.

21

Pisces is a constellation of the zodiac. Its vast bulk – and main asterism viewed in most European cultures per Greco-Roman antiquity as a distant pair of fishes connected by one cord each that join at an apex – are in the Northern celestial hemisphere. Its old astronomical symbol is Pisces symbol. Its name is Latin for "fishes". It is between Aquarius, of similar size, to the southwest and Aries, which is smaller, to the east. The ecliptic and the celestial equator intersect within this constellation and in Virgo. This means the sun passes directly overhead of the equator, on average, at approximately this point in the sky, at the March equinox.

22

Andromeda is one of the 48 constellations listed by the 2nd-century Greco-Roman astronomer Ptolemy, and one of the 88 modern constellations. Located in the northern celestial hemisphere, it is named for Andromeda, daughter of Cassiopeia, in the Greek myth, who was chained to a rock to be eaten by the sea monster Cetus. Andromeda is most prominent during autumn evenings in the Northern Hemisphere, along with several other constellations named for characters in the Perseus myth. Because of its northern declination, Andromeda is visible only north of 40° south latitude; for observers farther south, it lies below the horizon. It is one of the largest constellations, with an area of 722 square degrees. This is over 1,400 times the size of the full moon, 55% of the size of the largest constellation, Hydra, and over 10 times the size of the smallest constellation, Crux.

Its brightest star, Alpha Andromedae, is a binary star that has also been counted as a part of Pegasus, while Gamma Andromedae is a colorful binary and a popular target for amateur astronomers. Only marginally dimmer than Alpha, Beta Andromedae is a red giant, its color visible to the naked eye. The constellation's most obvious deep-sky object is the naked-eye Andromeda Galaxy (M31, also called the Great Galaxy of Andromeda), the closest spiral galaxy to the Milky Way and one of the brightest Messier objects. Several fainter galaxies, including

M31's companions M110 and M32, as well as the more distant NGC 891, lie within Andromeda. The Blue Snowball Nebula, a planetary nebula, is visible in a telescope as a blue circular object.

23

Triangulum is a small constellation in the northern sky. Its name is Latin for "triangle", derived from its three brightest stars, which form a long and narrow triangle. Known to the ancient Babylonians and Greeks, Triangulum was one of the 48 constellations listed by the 2nd century astronomer Ptolemy. The celestial cartographers Johann Bayer and John Flamsteed catalogued the constellation's stars, giving six of them Bayer designations.

The white stars Beta and Gamma Trianguli, of apparent magnitudes 3.00 and 4.00, respectively, form the base of the triangle and the yellow-white Alpha Trianguli, of magnitude 3.41, the apex. Iota Trianguli is a notable double star system, and there are three star systems with known planets located in Triangulum. The constellation contains several galaxies, the brightest and nearest of which is the Triangulum Galaxy or Messier 33—a member of the Local Group. The first quasar ever observed, 3C 48, also lies within the boundaries of Triangulum.

24

Aries is one of the constellations of the zodiac. It is located in the Northern celestial hemisphere between Pisces to the west and Taurus to the east. The name Aries is Latin for ram. Its old astronomical symbol is Aries symbol . It is one of the 48 constellations described by the 2nd century astronomer Ptolemy, and remains one of the 88 modern constellations. It is a mid-sized constellation ranking 39th in overall size, with an area of 441 square degrees (1.1% of the celestial sphere).

Aries has represented a ram since late Babylonian times. Before that, the stars of Aries formed a farmhand. Different cultures have incorporated the stars of Aries into different constellations including twin inspectors in China and a porpoise in the Marshall Islands. Aries is a relatively dim constellation, possessing only four bright stars: Hamal (Alpha Arietis, second magnitude), Sheratan (Beta Arietis, third magnitude), Mesarthim (Gamma Arietis, fourth magnitude), and 41 Arietis (also fourth magnitude). The few deep-sky objects within the constellation are quite faint and include several pairs of interacting galaxies. Several meteor showers appear to radiate from Aries, including the Daytime Arietids and the Epsilon Arietids.

25

Perseus is a constellation in the northern sky, being named after the Greek mythological hero Perseus. It is one of the 48 ancient constellations listed by the 2nd-century astronomer Ptolemy, and among the 88 modern constellations defined by the International Astronomical Union (IAU). It is located near several other constellations named after ancient Greek legends surrounding Perseus, including Andromeda to the west and Cassiopeia to the north. Perseus is also bordered by Aries and Taurus to the south, Auriga to the east, Camelopardalis to the north, and Triangulum to the west. Some star atlases during the early 19th century also depicted Perseus holding the disem-

Portal

bodied head of Medusa, whose asterism was named together as Perseus et Caput Medusae; however, this never came into popular usage.

The galactic plane of the Milky Way passes through Perseus, whose brightest star is the yellow-white supergiant Alpha Persei (also called Mirfak), which shines at magnitude 1.79. It and many of the surrounding stars are members of an open cluster known as the Alpha Persei Cluster. The best-known star, however, is Algol (Beta Persei), linked with ominous legends because of its variability, which is noticeable to the naked eye. Rather than being an intrinsically variable star, it is an eclipsing binary. Other notable star systems in Perseus include X Persei, a binary system containing a neutron star, and GK Persei, a nova that peaked at magnitude 0.2 in 1901. The Double Cluster, comprising two open clusters quite near each other in the sky, was known to the ancient Chinese. The constellation gives its name to the Perseus cluster (Abell 426), a massive galaxy cluster located 250 million light-years from Earth. It hosts the radiant of the annual Perseids meteor shower—one of the most prominent meteor showers in the sky.

26

Cetus (/'si?t?s/) is a constellation, sometimes called 'the whale' in English. The Cetus was a sea monster in Greek mythology which both Perseus and Heracles needed to slay. Cetus is in the region of the sky that contains other water-related constellations: Aquarius, Pisces and Eridanus.

27

Taurus (Latin for "the Bull") is one of the constellations of the zodiac and is located in the northern celestial hemisphere. Taurus is a large and prominent constellation in the Northern Hemisphere's winter sky. It is one of the oldest constellations, dating back to the Early Bronze Age at least, when it marked the location of the Sun during the spring equinox. Its importance to the agricultural calendar influenced various bull figures in the mythologies of Ancient Sumer, Akkad, Assyria, Babylon, Egypt, Greece, and Rome. Its old astronomical symbol is Taurus symbol, which resembles a bull's head.

A number of features exist that are of interest to astronomers. Taurus hosts two of the nearest open clusters to Earth, the Pleiades and the Hyades, both of which are visible to the naked eye. At first magnitude, the red giant Aldebaran is the brightest star in the constellation. In the northeast part of Taurus is Messier 1, more commonly known as the Crab Nebula, a supernova remnant containing a pulsar. One of the closest regions of active star formation, the Taurus-Auriga complex, crosses into the northern part of the constellation. The variable star T Tauri is the prototype of a class of pre-main-sequence stars.

28

Auriga is a constellation in the northern celestial hemisphere. It is one of the 88 modern constellations; it was among the 48 constellations listed by the 2nd-century astronomer Ptolemy. Its name is Latin for '(the) charioteer', associating it with various mythological beings, including Erichthonius and Myrtilus. Auriga is most prominent during winter evenings in the northern Hemisphere, as are five other constellations that have stars in the Winter Hexagon asterism. Because of its northern declination, Auriga is only visible in its entirety as far south as -34°; for observers farther south it lies partially or fully below the horizon. A large constellation, with an area of 657 square degrees, it is half the size of the largest, Hydra.

Its brightest star, Capella, is an unusual multiple star system among the brightest stars in the night sky. Beta Aurigae is an interesting variable star in the constellation; Epsilon Aurigae, a nearby eclipsing binary with an unusually long period, has been studied intensively. Because of its position near the winter Milky Way, Auriga has many bright open clusters in its borders, including M36, M37, and M38, popular targets for amateur astronomers. In addition, it has one prominent nebula, the Flaming Star Nebula, associated with the variable star AE Aurigae.

29

Eridanus (/?'r?d?n?s/) is a constellation in the southern celestial hemisphere. It is represented as a river. One of the 48 constellations listed by the 2nd century astronomer Ptolemy, it remains one of the 88 modern constellations. It is the sixth largest of the modern constellations, and the one that extends farthest in the sky from north to south. The same name was later taken as a Latin name for the real Po River and also for the name of a minor river in Athens.

30

Orion is a prominent constellation located on the celestial equator and visible throughout the world. It is one of the most conspicuous and recognizable constellations in the night sky. It is named after Orion, a hunter in Greek mythology. Its brightest stars are the blue-white Rigel (Beta Orionis) and the red Betelgeuse (Alpha Orionis).

31

Canis Minor /?ke?n?s 'ma?n?r/ is a small constellation in the northern celestial hemisphere. In the second century, it was included as an asterism, or pattern, of two stars in Ptolemy's 48 constellations, and it is counted among the 88 modern constellations. Its name is Latin for "lesser dog", in contrast to Canis Major, the "greater dog"; both figures are commonly represented as following the constellation of Orion the hunter.

Canis Minor contains only two stars brighter than the fourth magnitude, Procyon (Alpha Canis Minoris), with a magnitude of 0.34, and Gomeisa (Beta Canis Minoris), with a magnitude of 2.9. The constellation's dimmer stars were noted by Johann Bayer, who named eight stars including Alpha and Beta, and John Flamsteed, who numbered fourteen. Procyon is the eighth-brightest star in the night sky, as well as one of the closest. A yellow-white main-sequence star, it has a white dwarf companion. Gomeisa is a blue-white main-sequence star. Luyten's Star is a ninth-magnitude red dwarf and the Solar System's next closest stellar neighbour in the constellation after Procyon. Additionally, Procyon and Luyten's Star are only 1.12 light-years away from each other, and Procyon would be the brightest star in Luyten's Star's sky. The fourth-magnitude HD 66141, which has evolved into an orange giant towards the end of its life cycle, was discovered to have a planet in 2012. There are two faint deep-sky objects within the constellation's borders. The 11 Canis-Minorids are a meteor shower that can be seen in early December.

32

Monoceros (Greek: ?????e???, "unicorn") is a faint constellation on the celestial equator. Its definition is attributed to the 17th-century Dutch cartographer Petrus Plancius. It is bordered by Orion to the west, Gemini to the north, Canis Major to the south, and Hydra to the east. Other bordering constellations include Canis Minor, Lepus, and Puppis.

33

Gemini is one of the constellations of the zodiac and is located in the northern celestial hemisphere. It was one of the 48 constellations described by the 2nd century AD astronomer Ptolemy, and it remains one of the 88 modern constellations today. Its name is Latin for twins, and it is associated with the twins Castor and Pollux in Greek mythology. Its old astronomical symbol is Gemini

34

Hydra is the largest of the 88 modern constellations, measuring 1303 square degrees, and also the longest at over 100 degrees. Its southern end borders Libra and Centaurus and its northern end borders Cancer. It was included among the 48 constellations listed by the 2nd century astronomer Ptolemy. Commonly represented as a water snake, it straddles the celestial equator.

35

Lynx is a constellation named after the animal, usually observed in the Northern Celestial Hemisphere. The constellation was introduced in the late 17th century by Johannes Hevelius. It is a faint constellation, with its brightest stars forming a zigzag line. The orange giant Alpha Lyncis is the brightest star in the constellation, and the semi-regular variable star Y Lyncis is a target for amateur astronomers. Six star systems have been found to contain planets. Those of 6 Lyncis and HD 75898 were discovered by the Doppler method; those of XO-2, XO-4, XO-5 and WASP-13 were observed as they passed in front of the host star.

Within the constellation's borders lie NGC 2419, an unusually remote globular cluster; the galaxy NGC 2770, which has hosted three recent Type Ib supernovae; the distant quasar APM 08279+5255, whose light is magnified and split into multiple images by the gravitational lensing effect of a foreground galaxy; and the Lynx Supercluster, which was the most distant supercluster known at the time of its discovery in 1999.

36

Cancer is one of the twelve constellations of the zodiac and is located in the Northern celestial hemisphere. Its old astronomical symbol is Cancer symbol (fixed width).svg (??). Its name is Latin for crab and it is commonly represented as one. Cancer is a medium-size constellation with an area of 506 square degrees and its stars are rather faint, its brightest star Beta Cancri having an apparent magnitude of 3.5. It contains two stars with known planets, including 55 Cancri, which has five: one super-earth and four gas giants, one of which is in the habitable zone and as such has expected temperatures similar to Earth. At the (angular) heart of this sector of our celestial sphere is Praesepe (Messier 44), one of the closest open clusters to Earth and a popular target for amateur astronomers.

37

Sextans is a minor equatorial constellation which was introduced in 1687 by Johannes Hevelius. Its name is Latin for the astronomical sextant, an instrument that Hevelius made frequent use of in his observations.

38

Leo /'li?o?/ is one of the constellations of the zodiac, between Cancer the crab to the west and Virgo the maiden to the east. It is located in the Northern celestial hemisphere. Its name is Latin for lion, and to the ancient Greeks represented the Nemean Lion killed by the mythical Greek hero Heracles meaning 'Glory of Hera' (known to the ancient Romans as Hercules) as one of his twelve labors. Its old astronomical symbol is Leo symbol. One of the 48 constellations described by the 2nd-century astronomer Ptolemy, Leo remains one of the 88 modern constellations today, and one of the most easily recognizable due to its many bright stars and a distinctive shape that is reminiscent of the crouching lion it depicts. The lion's mane and shoulders also form an asterism known as "The Sickle," which to modern observers may resemble a backwards "question mark."

39

Leo Minor is a small and faint constellation in the northern celestial hemisphere. Its name is Latin for "the smaller lion", in contrast to Leo, the larger lion. It lies between the larger and more recognizable Ursa Major to the north and Leo to the south. Leo Minor was not regarded as a separate constellation by classical astronomers; it was designated by Johannes Hevelius in 1687.

There are 37 stars brighter than apparent magnitude 6.5 in the constellation; three are brighter than magnitude 4.5. 46 Leonis Minoris, an orange giant of magnitude 3.8, is located some 95 light-years from Earth. At magnitude 4.4, Beta Leonis Minoris is the second-brightest star and the only one in the constellation with a Bayer designation. It is a binary star, the brighter component of which is an orange giant and the fainter a yellow-white main sequence star. The third-brightest star is 21 Leonis Minoris, a rapidly rotating white main-sequence star of average magnitude 4.5. The constellation also includes two stars with planetary systems, two pairs of interacting galaxies, and the unique deep-sky object Hanny's Voorwerp.

40 My world, my own gate.

Bron Wikipedia

"Quantiversum Portal" BACK

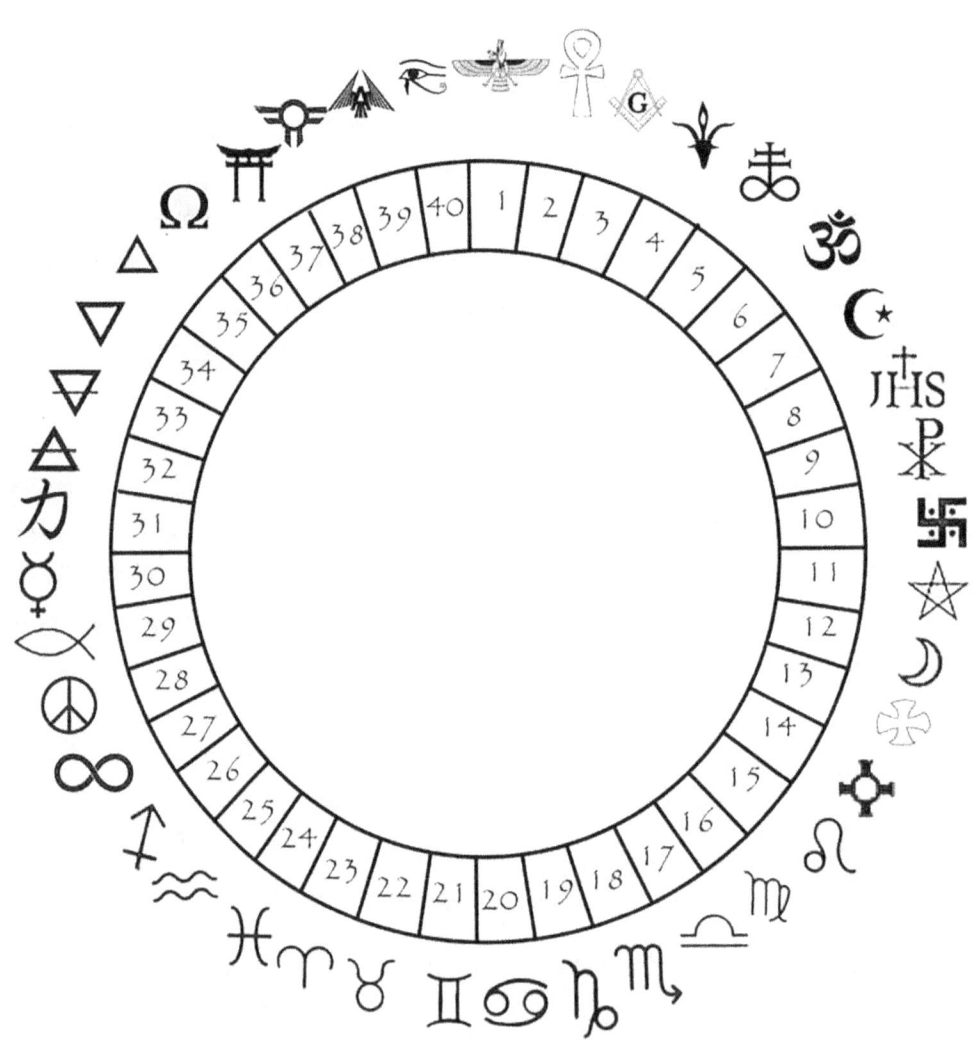

DESCRIPTION BACK

1 Ahura Mazda Faravahar

Is a prominent guardian spirit in Zoroastrianism and Iranian culture that is believed to be a depiction of a Fravashi.

2 The "Ankh" or "key of life"

Ancient Egyptian symbol for eternal life; now also associated with Kemetism and neo-paganism.

3 "Mason symbol Freemasonry"

Masonry refers to fraternal organisations that trace their origins to the local guilds of stonemasons that, from the end of the 13th century, regulated the qualifications of stonemasons and their interaction with authorities and clients.

4 "Capricorn" or "the devil"

Adopted by modern occultists and Satanists after the Knights Templar were accused of worshiping it. Theistic Satanists may worship it as a deity or demon, while atheistic Satanists see it as a metaphorical symbol.

5 Leviathan cross

Alchemical symbol for black sulfur, is also known as a "Leviathan Cross" or "Satan's Cross".

6 "Ohm" symbol "Omkara".

A sacred script for Hindus that celebrates "Brahman" or "God the Creator".

7 "Muslim symbol"

Islam is one of the three major religions that sprang forth from the Middle East, along with Judaism and Christianity.

It traces its origin back to Abraham, whom its adherents called Muslims believed worshipped Allah, their "one true God".

8 "Christian monogram" of Jesus Christ (Christogram).
The Savior, The Lord Our God.

Portal

9 Cross Constantine
This common symbol appears in various forms that represent Christianity or Christians. Its exact origin is unknown but the Christians appear to have adopted the symbol from the Greeks. The vertical line may have represented a cosmic tree and the axial symbol.

10 "Swastika Dalam Agama Hindu".
The swastika is a symbol with many styles and meanings and can be found in many cultures.

11 "Pentagram" (sometimes known as a pentalpha, pentangle, or star pentagon)
Is a regular five-pointed star polygon, formed from the diagonal line segments of a convex (or simple, or non-self-intersecting) regular pentagon. Pentagram symbols from about 5,000 years ago were found in the Liang-zhu culture of China.

12 Symbol Moon.

13 The secret world of templars.

14 Thought Symbol.

15 Constellation Leo.

16 Constellation Virgo.

17 Constellation libra.

18 Constellation Scorpio.

19 Constellation Capricorn.

20 Constellation Cancer.

21 Constellation Gemini.

22 Constellation Taurus.

23 Constellation Arius.

24 Constellation Pisces.

25 Constellation Aquarius.

26 Constellation Sagittarius.

27 The "infinity" symbol
I s a mathematical symbol representing the concept of infinity. This symbol is also called a lemniscate, after the lemniscate curves of a similar shape studied in algebraic geometry, or "lazy eight", in the terminology of livestock branding.

28 A "peace sign",
Which is widely associated with pacifism.

29 The "Ichthys" or "Ichthus" from the Greek ikhthýs.
AD Koine Greek "fish" is a symbol consisting of two intersecting arcs, the ends of the right side extending beyond the meeting point so as to resemble the profile of a fish. Now known colloquially as the "sign of the fish" or the "Jesus fish".

30 Crossed caduceus symbol for Mercury.
The symbol for Mercury is a caduceus (a staff intertwined with two serpents), a symbol associated with Mercury/Hermes throughout antiquity. Sometime after the 11th century, a cross was added to the bottom of the staff to make it seem more Christia.

31 Japanese "Katakana kyokashotai"
KA, Symbol for Strength.

32 Alchemy Element air.

33 Alchemy Element earth.

34 Alchemy Element water.

35 Alchemy Element fire.

36 Omega
W as not part of the early (8th century BC) Greek alphabets. It was introduced in the late 7th century BC in the Ionian cities of Asia.

37 Japanese Buddhist "Torii gate".

38 Symbol Niburu.

39 Symbol Anunnaki.

40 The eye of the "god Horus".
Symbol of protection, now associated with the occult and Kemetism.

Frequency
Quativersum Portal

These books can be ordered on the website;
http://www.johnbaselmans.com/Books/Books.htm
The published books are:

John Baselmans Drawing Course	ISBN 978-0-557-01154-4
The secrets behind my drawings	ISBN 978-0-557-01156-8
The world of human drawings	ISBN 978-0-557-02754-5
Drawing humans in black and white	ISBN 978-1-4092-5186-6
John Baselmans' Lifework part 1	ISBN 978-1-4092-8941-8
John Baselmans' Lifework part 2	ISBN 978-1-4092-8959-3
John Baselmans' Lifework part 3	ISBN 978-1-4092-8974-6
John Baselmans' Lifework part 4	ISBN 978-1-4092-8937-1
John Baselmans' Lifework de luxe part 1	
John Baselmans' Lifework de luxe part 2	
John Baselmans' Lifework de luxe Curriculum	
Eiland-je bewoner Deel 1	ISBN 978-1-4092-1856-2
Eiland-je bewoner Deel 2	
Eilandje bewoner - Luxe edition	ISBN 978-1-4092-2102-9
Eiland-je bewoner Bundel	ISBN 978-0-557-01281-7
Mañan	ISBN 978-1-4092-8949-4
He oudje leef je nog?	ISBN 978-1-4092-8482-6
De wijsheden van onze oudjes	ISBN 978-1-4092-9516-7
Makamba	ISBN 978-1-4461-3036-0
Onze Cultuur	ISBN 978-1-4475-2701-5

Ingezonden	ISBN 978-1-4092-1936-1
Moderne slavernij in het systeem	ISBN 978-1-4092-5909-1
Help, de Antillen verzuipen	ISBN 978-1-4092-7972-3
Geboren voor één cent	ISBN 978-1-4452-6787-6
Pech gehad	ISBN 978-1-4457-6170-1
Zwartboek	ISBN 978-1-4461-8058-7
Mi bida no bal niun sèn	ISBN 978-1-4467-2954-0
Curacao Maffia Eiland	ISBN 978-1-4478-9049-2
De missende link	ISBN 978-1-4710-9498-9
Curatele	ISBN 978-1-4717-9319-6
Curacao achter gesloten deuren	ISBN 978-1-304-58901-9
De MATRIX van het systeem deel 1	ISBN 978-1-291-88840-9
De MATRIX van het systeem deel 2	ISBN 978-1-291-88841-6
The hidden world part 1	ISBN 978-1-326-03644-7
The hidden world part 2	ISBN 978-1-326-03645-4
Geloof en het geloven	ISBN 978-1-326-28453-4
Dieptepunt	ISBN 978-1-326-71278-5
Namen / Names	ISBN 978-1-326-81898-2
Drugs	ISBN 978-1-326-84325-0
De protocollen van Sion 21ste eeuw	ISBN 978-0-244-61655-7
Verboden publicaties	ISBN 978-0-244-91960-3
De maatschappelijke beerput	ISBN 978-0-244-36559-2
Achter de sociale media schermen	ISBN 978-0-244-14015-1
Project Corona/ COVID-19	ISBN 978-1-71664-848-9
Omnis 1	ISBN 978-0-244-10848-9
Omnis 2	ISBN 978-0-244-40848-0
Omnis 3	ISBN 978-0-244-70848-1
Omnis 4	ISBN 978-0-244-10849-6
Omnis 5	ISBN 978-0-244-40849-7
Omnis 6	ISBN 978-0-244-81855-5
Omnis 7	ISBN 978-1-716-67541-6
Omnis 8	ISBN 978-1-008-91377-6
Omnis 9	ISBN 978-1-716-00185-7
Omnis Photos	ISBN 978-0-244-10859-5
Omnis Photos 2	ISBN 978-0-244-44015-2
Omnis Photos 3	ISBN 978-0-244-21863-8
Omnis Photos 4	ISBN 978-1-71664-569-3
Omnis Photos 5	ISBN 978-1-71664-567-9
Omnis Photos 6	ISBN 978-1-008-91378-3
Omnis Photos 7	ISBN 978-1-008-91375-2
Omnis Photos 8	ISBN 978-1-716-00187-1

Portal

Words of wisdom (part 1)	ISBN 978-1-4452-6789-0
Words of wisdom (part 2)	ISBN 978-1-4452-6791-3
Words of wisdom (part 3)	ISBN 978-1-4461-3035-3
Words of wisdom (part 4)	ISBN 978-1-4710-8130-9
The world of positive energy	ISBN 978-0-557-02542-8
Het energieniale leven	ISBN 978-1-4457-2953-4
Dood is dood "energieniale leven"	ISBN 978-1-4476-7213-5
Zelfgenezing "energieniale leven"	ISBN 978-1-4709-3332-6
Levenscirkel "energieniale leven"	ISBN:978-1-300-76189-1
Utopia "energieniale leven"	ISBN 978-1-329-51188-0
Vrijheid en liefde "energieniale leven"	ISBN 978-1-329-79390-3
Dimensies "energieniale leven"	ISBN 978-1-365-87087-3
Hologram "energieniale leven"	ISBN 978-1-387-72155-9
Het lang verborgen geheim "energieniale leven"	ISBN 978-0-359-70533-7
Quantiversum "energieniale leven"	ISBN 978-1-71657-634-8
The Architect "energieniale leven"	ISBN 978-1-7947-4982-5
NU deel 1	ISBN 978-1-4092-7691-3
NU deel 2	ISBN 978-1-4092-7736-1
NU deel 3	ISBN 978-1-4092-7747-7
NU deel 4	ISBN 978-1-4092-7787-3
NU deel 5	ISBN 978-1-4092-7720-0
NU deel 6	ISBN 978-1-4092-7742-2
NU deel 7	ISBN 978-1-4092-7775-0
NU deel 8	ISBN 978-1-4092-7738-5
NU deel 9	ISBN 978-1-4092-7768-2
NU deel 10	ISBN 978-1-4092-7708-8
NU deel 11	ISBN 978-1-4092-7759-0
NU deel 12	ISBN 978-1-4092-7661-6

www.ingramcontent.com/pod-product-compliance
Lightning Source LLC
Chambersburg PA
CBHW081013170526
45158CB00010B/3027

9 781312 549319